Amazing Animals

Spider Monkeys

Place Value

Logan Avery

Consultants

Colene Van Brunt
Math Coach
Hillsborough County Public Schools

Publishing Credits
Rachelle Cracchiolo, M.S.Ed., *Publisher*
Conni Medina, M.A.Ed., *Managing Editor*
Dona Herweck Rice, *Series Developer*
Emily R. Smith, M.A.Ed., *Series Developer*
Diana Kenney, M.A.Ed., NBCT, *Content Director*
June Kikuchi, *Content Director*
Susan Daddis, M.A.Ed., *Editor*
Karen Malaska, M.Ed., *Editor*
Kevin Panter, *Senior Graphic Designer*

Image Credits: p.11 Lelia Valduga/Getty Images; all other images from iStock and/or Shutterstock.

Library of Congress Cataloging-in-Publication Data

Names: Avery, Logan, author.
Title: Spider monkeys / Logan Avery.
Description: Huntington Beach, CA : Teacher Created Materials, Inc., [2018] | Series: Amazing animals | Audience: K to grade 3. | Includes index. |
Identifiers: LCCN 2017054933 (print) | LCCN 2017056330 (ebook) | ISBN 9781480759725 (eBook) | ISBN 9781425856786 (pbk.)
Subjects: LCSH: Spider monkeys--Juvenile literature.
Classification: LCC QL737.P915 (ebook) | LCC QL737.P915 A94 2018 (print) | DDC 599.8/58--dc23
LC record available at https://lccn.loc.gov/2017054933

Teacher Created Materials

5301 Oceanus Drive
Huntington Beach, CA 92649-1030
www.tcmpub.com

ISBN 978-1-4258-5678-6
© 2019 Teacher Created Materials, Inc.

Table of Contents

Monkey Business 4

Hands and Tails 10

Monkey Around 16

Problem Solving 20

Glossary 22

Index 23

Answer Key 24

Monkey Business

Is it a spider? Is it a monkey? It is a spider monkey!

Spider monkeys come in different colors and sizes.

Spider monkeys have long, thin legs. They like to hang upside down. It makes them look like spiders. That is how they got their name.

But these monkeys are not spiders. In fact, they eat spiders! They also eat a lot of fruit and parts of plants.

This spider monkey chews a piece of grass.

LET'S DO MATH!

A spider monkey eats 30 seeds. Use straws to stand for seeds. Make bundles of 10.

1. How many bundles of ten are there?

2. How many leftover ones are there?

This spider monkey eats fruit.

Hands and Tails

Spider monkeys have hands shaped like **hooks**. They help them **grab** branches and **swing** through trees.

A spider monkey has no thumbs.

A young spider monkey uses its tail to hang from a tree.

These small monkeys can use their tails to grab and swing. Their tails work like hands.

Let's Do Math!

There are 50 spider monkeys hanging around. Use straws to stand for tails. Make bundles of 10. Write how many tens and leftover ones there are on the place value chart.

tens	ones

Spider monkeys can run, **climb**, and swing very fast. They move much faster than people can.

This spider monkey can hang and eat fruit at the same time.

This spider monkey uses its tail for balance.

Monkey Around

Spider monkeys live in large groups. They hug each other with their **limbs** and tails.

Three spider monkeys sit together on a tree.

A family looks for food.

LET'S DO MATH!

A **troop** of 60 brown spider monkeys swings through the trees. A troop of 40 golden spider monkeys naps in the shade. A troop of 70 red spider monkeys eats spiders. Use <, >, or = to compare the troops.

1. 60 _____ 40

2. 40 _____ 70

3. 70 _____ 60

Spider monkeys live in trees. They spend little time on the ground.

Maybe treetops are the best places to **monkey around**!

Problem Solving

Pretend you see spider monkeys in the jungle. Find out how many spider monkeys there are of each kind. Then, compare the numbers with words and symbols.

1. The number of _____ spider monkeys is greater than the number of _____ spider monkeys.

   _____ > _____

2. The number of _____ spider monkeys is less than the number of _____ spider monkeys.

   _____ < _____

Kind of Spider Monkeys	Description	Number
brown	7 tens, 0 ones	____
red	____ tens, ____ ones	20
black	____ tens, ____ ones	50

Glossary

climb—to use hands and feet to move up or down a large object

grab—to take quickly in hand and hold onto

hooks—tools that are curved and are used to catch or hold things

limbs—arms, legs, or tails

monkey around—to play instead of work

swing—to go forward and backward while hanging from something

troop—a group of spider monkeys

Index

groups, 16

hands, 10, 13

limbs, 16

tails, 12–13, 15–16

Answer Key

Let's Do Math!

page 9:

1. 3

2. 0

page 13:

tens	ones
5	0

page 17:

1. >

2. <

3. >

Problem Solving

brown—70;
red—2, 0;
black—5, 0

1. Answers will vary.

   Example: brown, red; 70 > 20

2. Answers will vary.

   Example: red, black; 20 < 50